Notes

True Round Metal Boat Design

by way of
Approximate Development

D L Schaffer

The Design procedures detailed in this book reflect the authors' own experience and research. The authors takes no responsibility for the use or misuse of the information contained herein.

Copyright - 2026

No part of this book may be reproduced, stored in a retrieval system, or transmitted, in any form or by any means, electronic, mechanical, photocopying, recording, or otherwise, without the written permission of the publisher.

Introduction

True Round Metal Boat Design - By way of Approximate Development combines todays three-dimensional **Surface** design software with Approximate Development, which is a layout technique that divides a warped – True Round – Compound surfaces into segments using mathematical calculations onto a single plane for fabrication.

The **Surface** based software that I will used in this book is - Multi-surf by Aerohydro. Other **Surface** software design software will work equally well. Rhino comes to mind.

While I am going to jump right in with the design process, you probably should review the below subject's at the end of the book, before beginning.

- The 'Theory of 'Approximate Development'
- The ' Hierarchy of Entities'
- Other hull configurations
- The control points used to create the Model
- You may also find that reading my other book - ***True Round Metal Boat-Building – By way of Approximate Development***, written for 'Builder', helpful.

D L Schaffer

Index

The Hull Surface

Defining the C-lofted surface:	Page 7
Static Hydrostatics:	Page 9

The Framing System

Transverse Frames:	Page 11
Longintudinal Frames:	Page 13
Longintudinal Frame Surfaces:	Page 14
Relationship between Frames:	Page 16

True Round Shell Plating

True Round Shell Plating:	Page 19
Shell Plate – Frame relationship:	Page 22

Appendix's

Other Surface Configurations:	Page 27
Theory of Approximate Development:	Page 35
Hierarchy of Entities:	Page 51
Ruled vs Developable Surfaces:	Page 56
Master Curve Points:	Page 59
Bezier 2.5 Photo's:	Page 69

Designing
The Bezier 12.5

(A Single surface design)

Creating the C-Lofted Hull Surface:

Our first task is to define the hull. To that end all the control points that define the hull are given in an appendix at the end of the book. Therefore, you first task is to use th0se points to create a model file in whatever **Surface** design software you are using.

The prospective drawing below illustrates all the absolute points that define the seven (7) type 2 B-spline curves master curves that define the shape of the hull surface.

While, using the same point to define the (4) longintudinal C-spline curves which run the length of the hull which are essentially used to fair the hull. These curves are essential to fairing the hull. They are created by a type three C- spline entity where the curve passes through all the control points. The following drawing is a prospective view of the above input points and curves. I add the keel surface to the drawing to provide a little dimension to the drawing.

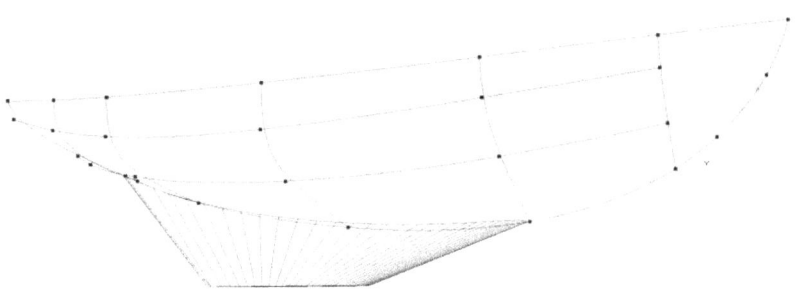

The surface shown below was created using a C-Lofted Surface entity 'parented' by the seven (7) transverse master curves defined above. I included the surface of the keel to add a little dimension to the illustration.

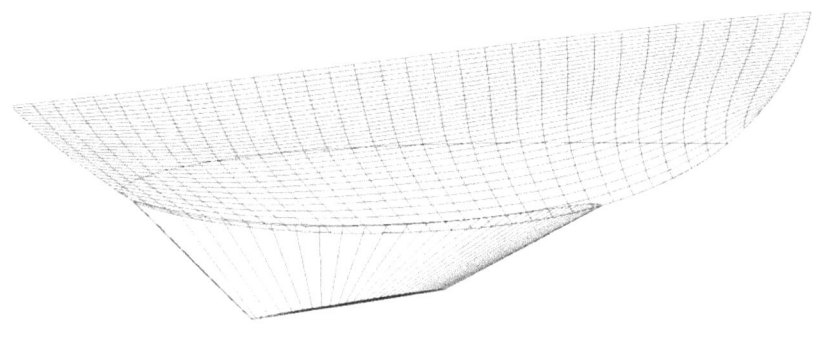

Static Hydrostatic:

While not directly related to the building process I thought the Reader might be interested in the Static Hydrostatic of the design. This is done in Multi-surf, by using the entity 'Counter' to divide the waterline length of the hull into ten (10) equal spaces as illustrated in the following sketch.

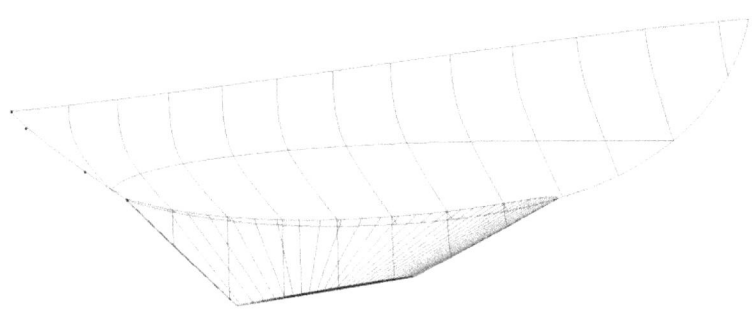

```
Sink                0.00 in          Spec. Wt.       0.03705 lb/in^3
Trim, deg.          0.00             Z c.g.             0.00 in
Heel, deg.          0.00
      Dimensions
W.L. Length       147.40 in          W.L. Beam         60.19 in
W.L. Fwd. X        20.39 in          Draft             29.17 in
W.L. Aft  X       167.78 in
      Displacement
Volume          42037.9 in^3         Ctr.Buoy. X      100.81 in
Displ't.         1557.5 lb           Ctr.Buoy. Y        0.00 in
LCB (% w.l.)       54.6              Ctr.Buoy. Z       -5.92 in
      Waterplane
W.P. Area        5934.22 in^2        Ctr.Flotn. X     101.98 in
LCF (% w.l.)       55.4
      Wetted Surface
Wetd.Area        9120.24 in^2        Ctr. W.S. X      104.21 in
Ctr. W.S. Z       -9.96 in
      Lateral Plane
L.P. Area        2732.16 in^2        Ctr. L.P. X      102.76 in
Ctr. L.P. Z      -11.83 in
      Initial Stability
Trans. GM         24.46 in           Trans.RMPD       664.8 in-lb
Longl. GM        150.39 in           Longl.RMPD      4087.6 in-lb
      Coefficients
Waterplane         0.669
Prismatic          0.563
Block              0.162
Midsection         0.289
Disp/length      375.1
```

The Framing System

The other requirement you need to be aware of is that the framing system needs to be Longitudinal. That is, where widely spaced transverse frames are supported by closely spaced longitudinal frames, which in turn support the shell plating. This framing system is essential to the design that I am about to describe.

Transverse Frames:

I am sure you can do your own calculations but my calculation provide for eight (8) transverse frames located at the below 'x' positions.

To profile a Transverse Frame onto the surface of the hull use the 'counter' entity. The 'Parents' needed to create a 'counter' is are location along the 'x' axis on the hull and the surface on which it lies.

Transverse frame_#1	'X' Position 14.500
Transverse frame_#2	'X' Position 36.750
Transverse frame_#3 `	'X' Position 59.000
Transverse frame_#4	'X' Position 81.250
Transverse frame_#5	'X' Position 103.500
Transverse frame_#6	'X' Position 125.75
Transverse frame_#7	'X' Position 148.000
Transverse frame_#8	'X' Position 165.750

The below illustration shows the profile created for Transverse frame #4 in the 'y' axis.

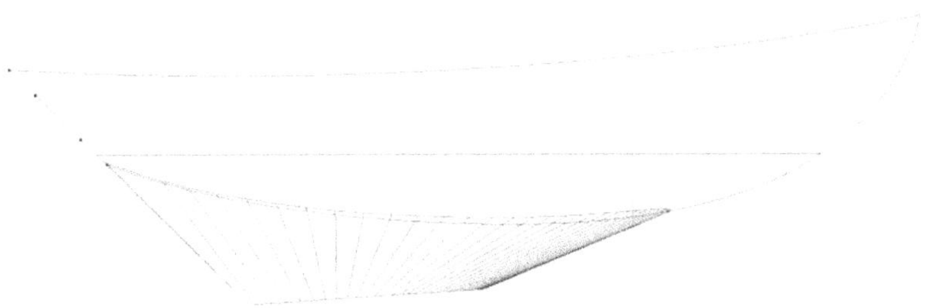

The below illustration the counter created at its 'x' location of 81.250" aft of the Fore Point of the hull.

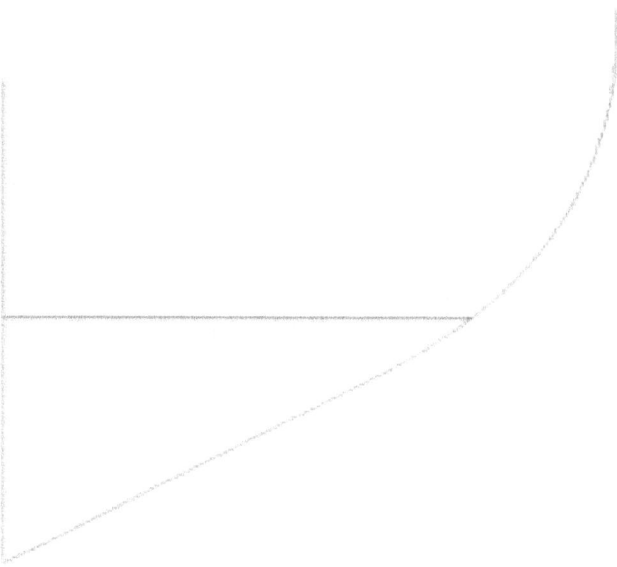

Longitudinal Frames:

The Bezier 12.5 has seven (7) longitudinal frames on each side of the hull. Their location on the Girth of the hull is defined by Absolute Magnets. Insert all the following 'magnets' using the, c-lofted surface, (hull-bezier-surf) as the 'parent' surface into your Model File. A magnet, in Mult-surf, is a entity that lies of a surface, and is defined by 'U' and 'V') location on this surface.

Long_1_magnet	'u' – 0.176	'v' – 0.471
long_2_magnet	'u' – 0.381	'v' – 0.472
long_3_magnet	'u' – 0.581	'v' – 0.467
long_4_magnet	'u' – 0.703	'v' – 0.462
long_5_magnet	'u' – 0.748	'v' – 0.461
long_6_magnet	'u' – 0.874	'v' – 0.460
long_7_magnet	'u' – 0.941	'v' – 0.460

Next, insert the following 'uvsnake' into the Model File. They will be 'parented' by the previous magnets. A 'uv-snake' in Multi-surf, is confined to lie on a surface and is located the uv direction of the surface by its above 'parent' magnet.

Long_1_uv	Parent = long_1_magnet
long_2_uv	Parent = long_2_magnet
long_3_uv	Parent = long_3_magnet
long_4_uv	Parent = long_4_magnet
long_5_uv	Parent = long_5_magnet
long_6_uv	Parent = long_6_magnet
long_7_uv	Parent = long_7_magnet

The prospective drawing. Below, reflects the 'magnets' and 'uv-snakes' on the hulls surface, C-spline lofted surface.

The next drawing is a profile view, with attached keel, shows the location and crossing between the Transverse and Longitudinal frames.

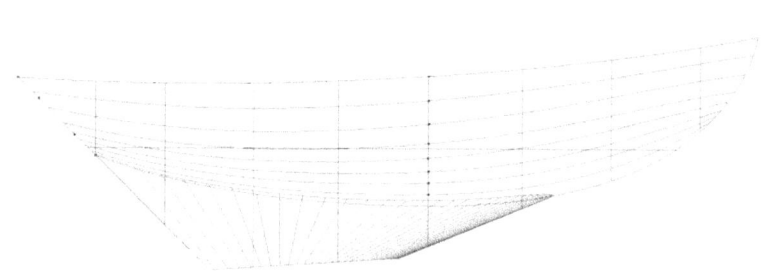

Longitudinal Frame Surfaces:

The longintudinal frames be fabricated from 1.000" x 0.250" flat bar.

The next step is to create an 'Offset Curve' parented by the UV-snakes previously created to define the location of each longintudinal frame.

(An 'Offset Curve' is a, 'Multi-surf entity', that takes each point on the snake and creates another curve the offset distance along the normal to the surface. Offset1 and Offset2 are signed decimal numbers which specify the offset distance at the t=0 and t=1 ends of parent Snake.)

Since the parent snake lies of the outside surface of the hull whose thickness is 0.125", therefore the Offset dimension at both ends of the Offset curve will be 0.125" + 1.000" the width of the framing. Offset distance therefore will be 1.125". Create the following 'Offset Curves.

'offset curve'	long_ s_ width
'offset curve'	long_1_width
'offset curve'	long_2_width

'offset curve'	long_3_width
'offset curve'	long_4_width
'offset curve'	long_5_width
'offset curve'	long_6_width
'offset curve'	long_7_width

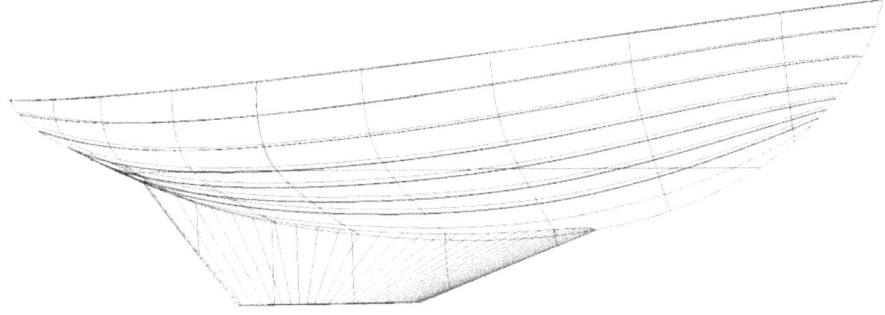

Next, the preceeding 'uv snake' and 'offset curves' are used to 'Parent' a ruled surface to define the longintudinal Frame surface

'ruled surf'	ruled_long_surf_s
'ruled surf'	ruled_long_surf_1
'ruled surf'	ruled_long_surf_2
'ruled surf'	ruled_long_surf_3
'ruled surf'	ruled_long_surf_4
'ruled surf'	ruled_long_surf_5
'ruled surf'	ruled_long_surf_6
'ruled surf'	ruled_long_surf_7

The following illustrations the above created hull longintudinal – surfaces in a prospective view. Your model file should now look similar to this illustration. Note: Go to Appendix on Ruled surfaces – Developable surfaces for insight on when to use a Ruled surface instead of a Developable surface.

The Longitudinal – Transverse Frame Relationship:

The illustration below shows the **'ruled'** surface that defined the Sheer Longitudinal un-folded onto a single plane using the Muli-surf utility program 'Msdev'. Note, that the unfolding process also locates the crossing of the transverse frames.

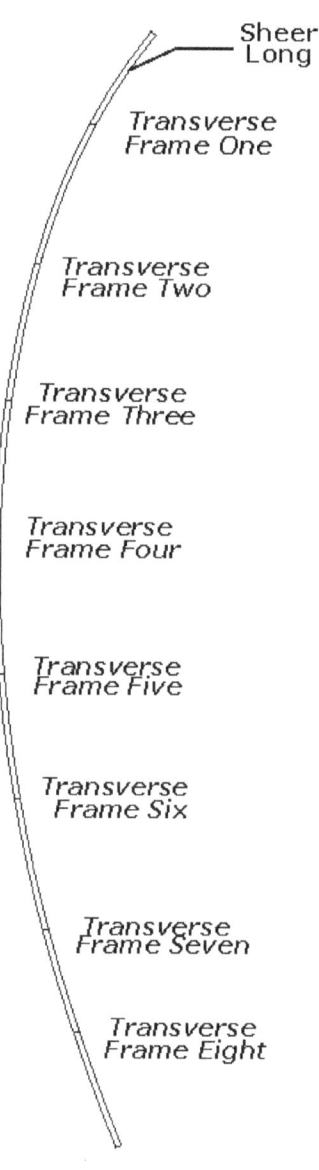

The Transverse - Longitudinal Frame Relationship:

Now that the Longitudinal frames have been defined, we can locate their crossing on the transverse frames.

To do this go back to the 'counters' entity used to profile the transverse frames onto the 'c-spline lofted Surface ' – hull-bezier and add the aforementioned Longitudinal frame surfaces to the list of surfaces at each Trans frame 'counter', shown below.

'counter' – frame_one'	'X' Position 14.500
'counter' – frame_two'	'X' Position 36.750
'counter' – frame_three'	'X' Position 59.000
'counter' – frame_four'	'X' Position 81.250
'counter' – frame_five'	'X' Position 103.500
'counter' – frame_six'	'X' Position 125.75
'counter' – frame_seven'	'X' Position 148.000
'counter' – frame_eight'	'X' Position 165.750

This X-plane profile view of Transverse frame #4 and its longitudinal frame crossing was isolated in Multi-surf and saved as a 2d DXF file for import into a 2D drafting where other details of the frame can be added.

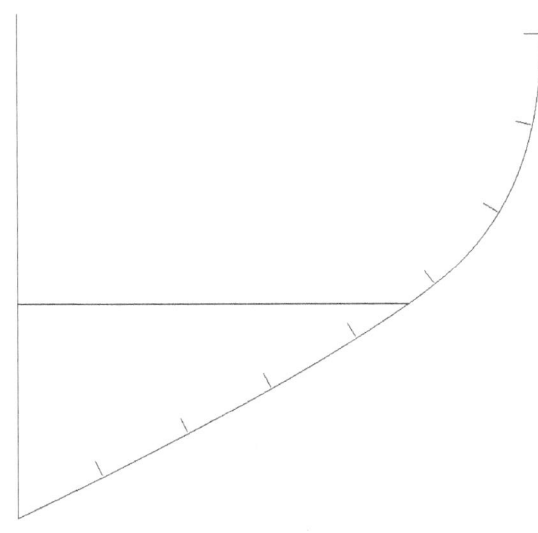

Before approximately developed Shell Plating lets take a break. The following photos show what our three-dimensional model file would look like in the real-world .

True Round Shell Plating

To create the approximately developed segmented True Round shell plates, 'line snakes' parented 'Magnets' on the base C-lofted surface are need along the sheer-line and fairbody lines, at intervals that will be used to divide the base C-lofted surface, as described in the appendix **'Theory of Approximate development'**.

The following drawing illustrates such a division created by 'magnets' along the sheer-line and fairbody lines.

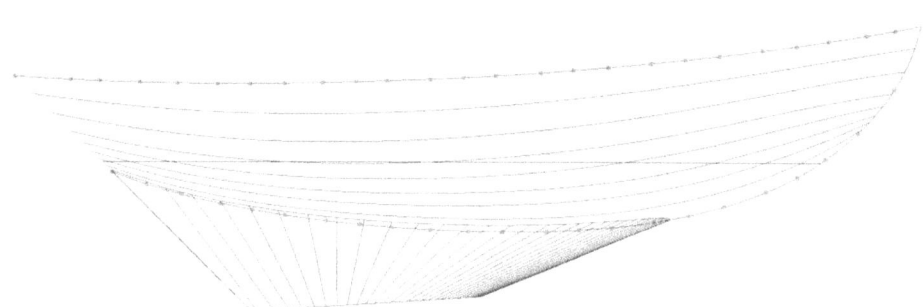

The next illustration shows the 'line snakes' that are 'parented' by the proceeding created 'magnets.

(Note that, the 'line-snakes' entity represents the shortest distance of a curved line between two points).

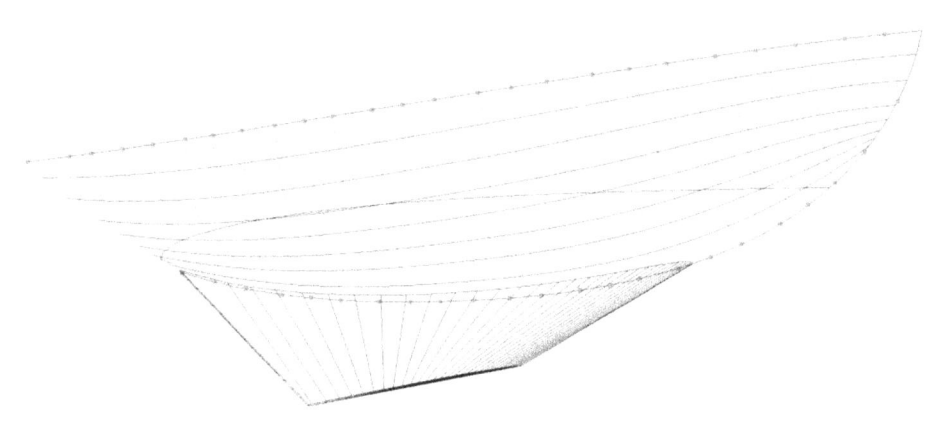

The 'developable surface', show below was 'parented' by the two previously created 'Line snakes' in way of Transverse frame #4 denoted by the curved line near the center of the surface represented by thee askew, unevenly spaced lines. More on these lines to follow.

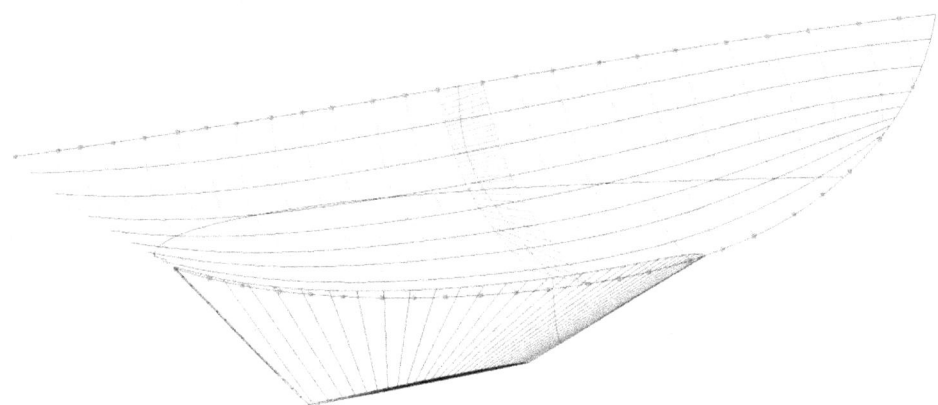

In the next drawing a Developable has been inserted between all the 'Line snakes'

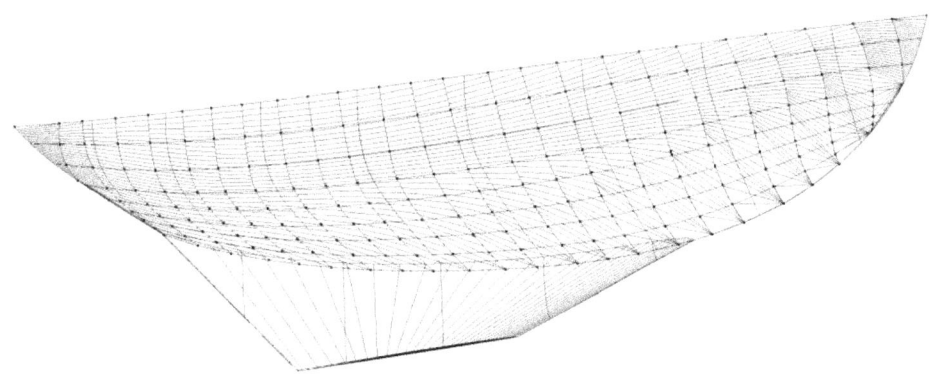

Shell Plate - Longitudinal Intersections:

So far, we have established the following intersections:

- The Transverse frames and shell plating:
- The Transverse frames and longitudinal frames.

The last intersection that needs to be defined is the crossing between the longitudinal frames and the approximately developed segmented True Rond shell plating.

To my dismay, as far as I can see, Multi-surf does not have an Entity that can establish a 'magnet' on a surface between two snakes that lie on that same surface.

My work around was to place 'magnet' on the approximately developed segment True Round shell plate 'By eye'. It is a bit of a task, but not unreasonable so.

The below illustration is a screen shoot of the approximately developed True Round segmented surfaces in way of Frame #4 where 'magnets. 'parented' by the developable surface True Round segment shell plate placed by Eye where the longitudinal's cross the edge of the surface.

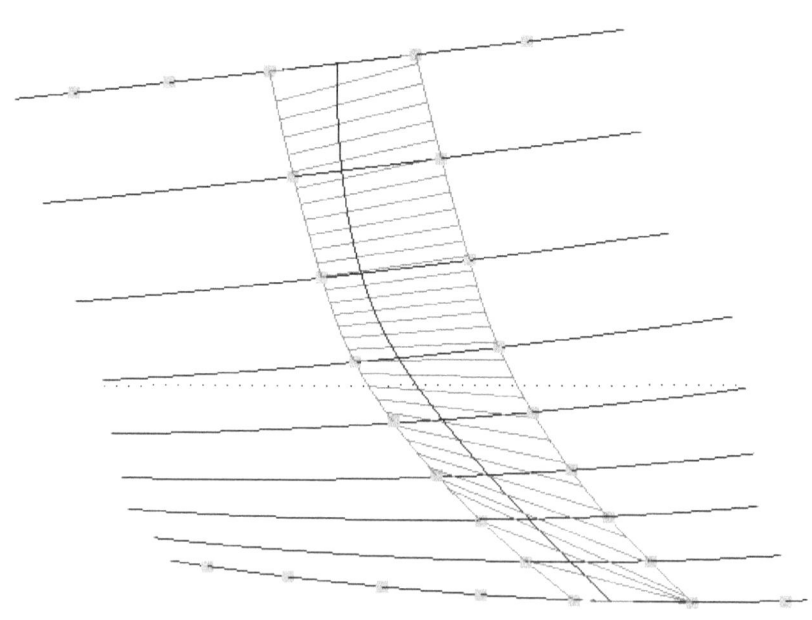

Unfolding the Shell Plating:

The shell plate in way of Frame #4 was unfolded, next page, was unfolded onto a single plane using Multi-surf utility software program Msdev. Msdev, provides a wealth of information needed to form the unfolded pattern back to its intended three-dimensional shape.

- The circle at the upper left short edge indicates that this edge is located at the sheer-line

- The single small half circles along this short sides of the unfolded pattern align the shell plating to Transverse Frame #4.

- The curved vertical line, near the center of the pattern represents where Transverse Frame #4 is in contact with the shell plating. It is interesting to note that the curved line will become straight after the pattern has been formed.

- In this particular pattern there are twenty-three (23) 'Element' lines. These lines indicate where the 'Press-break' tooling will engage the material to form the pattern back to its three-dimensional shape. The bend angle chosen for all shell patterns, in this design, is three (3) degrees.

- The small half circles along the vertical edges of the pattern indicate the locations where the shell plating crosses a longitudinal frame.

- The dimensioned cross lines (48.314" and 47.504) that run diagonally corner to corner are the of the shell plate after it has been formed back to its three-dimension shape.

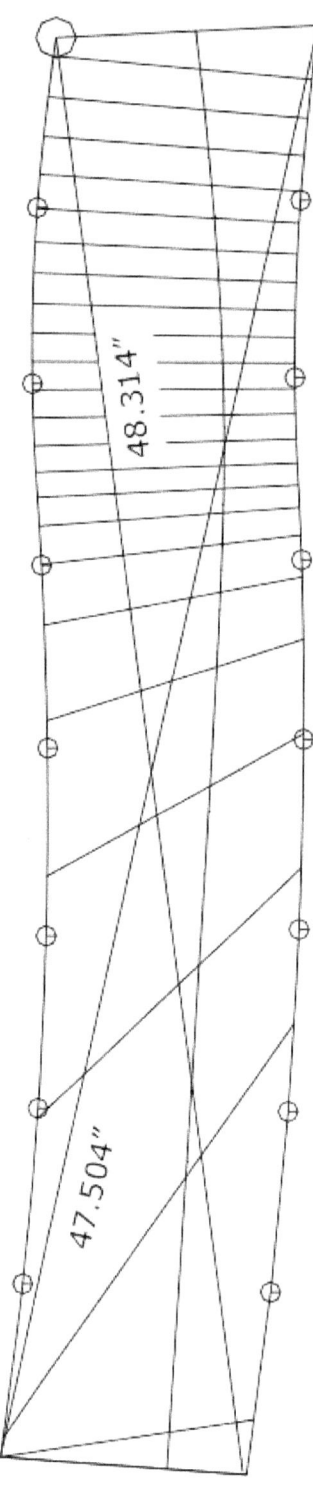

Other elements to note about the unfolded Pattern:

- Notice that the pattern runs the entire Girth of the hull from Sheerline to Fairbody Line.

- Notice that the three (3) degree 'Bend Lines' are not parallel to each other.

- Notice that the space between the 'Bend Lines' vary in distance.

- Notice that the 'Bend Lines' not only vary in distance, but in distance end to end.

- Notice the overall corner to corner measurements for the unfolded pattern

- Notice the curved line that locates Transverse Frame #4. The curved line will become straight after forming to align with the Transverse frame.

Cross Referencing:

After the segment shell plate section has been formed, it can be verified for accuracy by referencing to the cross-checking dimensions.

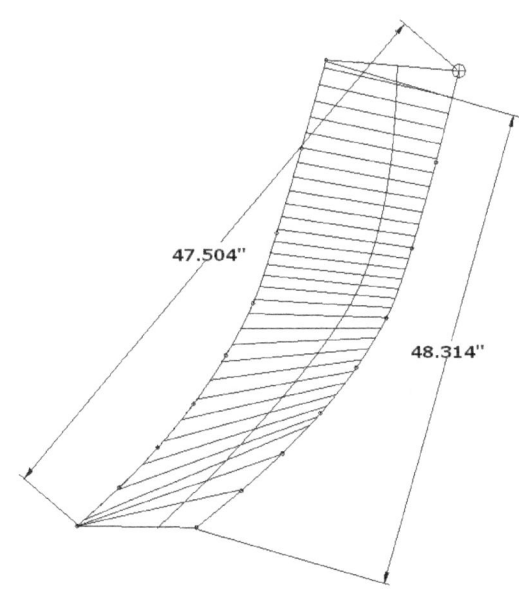

The finial result of our work is shown below:

- The hull is fair
- The Design objective have been met.
- The Transverse frames are defined.
- The Longintudinal frames are defined.
- The True Round Shell plates are defined.
- The intersection between all the hull components align.
- This hull can be assembled like a Childs Erector Set'.

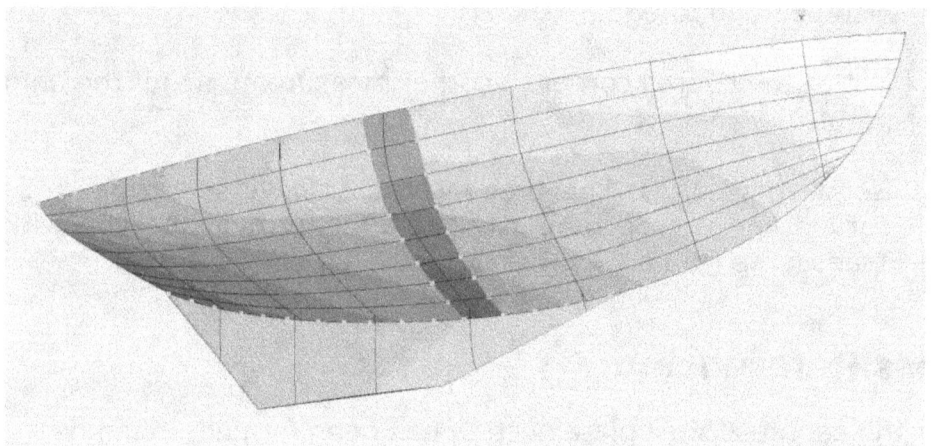

Congratulation

You have made it to the end of the design process. There is no more, unless you want to try out the other hull configuration in shown in the next chapter.

The next step is to unfold all the surfaces onto a single plane and create the two-dimensional drawing required by Builders. However, my other book, ***True Round Metal Boatbuilding – by Way of Approximate Development,*** may provide a little guidance.

or

Visit my website – metalsilboats.com

Other Hull Surface Configurations

Single True Round surface hull design:

The forgoing tutorial defined a hull entirely by a single warped surface, that ran from the sheerline around the turn of the bilge to the fairbody line at the bottom center of the hull be it a canoe body hull with attached keel or a wineglass hull form. You can look at a 'Bezier Single Surface' design being akin to any true round fiberglass sailboat hull design. The single surface hull form is the least complex from both a design and construction standpoint. Below is the section Lines view of the single surfaced 'Bezier 12.5.

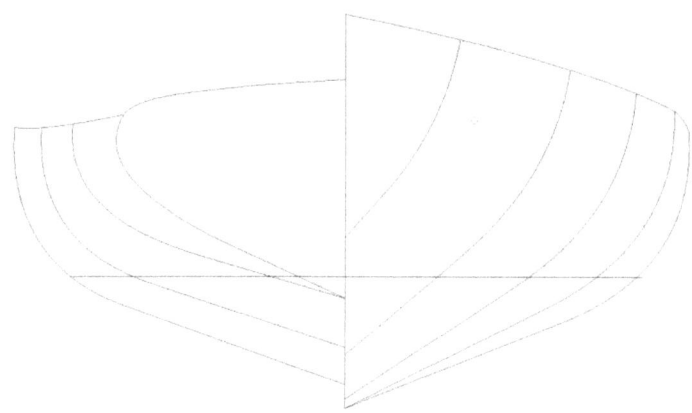

Preference for a 'Single Surface' Hull Configuration:

Since building the 'Double Surfaced' Bezier 12.5, I have come to the realization that combining a True Round surface with a Developable bottom surface serves no real purpose.

I now feel, that a 'Single' surface True Round design configuration has many advantages over the double or triple surface hull configurations that I was compelled to redesigned the double surface prototype version of the 'Bezier 12.5' to a 'Single' surface configuration.

The advantages of a 'Single' Surface Design over a 'Double' Surface design are:

- The Design Process itself was significantly simplified with the absence of a Theoretical chine.
- The longitudinal framing system is less problematic.
- Welding is simplified – No longitudinal shell plate weld seams.
- A single plating method.

In the redesign of the 'Bezier 12.5 from a 'Double' to a 'Single' hull configuration several hull features were strictly retained.

- Shape and position of the Sheer-line.
- Shape and position of the Fairbody line.
- Location and shape of the Bow.
- Position, shape, and angle of the transom.
- Location and configuration of the Keel and Rudder.

With the forgoing changes there was little or no change in Displacement, Center of Buoyancy, Center of Gravity, Stability, and Design Coefficients.

The following two Illustrations represent the 'Master Curves' that define both versions of the Bezier 12.5 overlayed onto each other. It can be seen that there is little difference in hull shape between the two hull configurations.

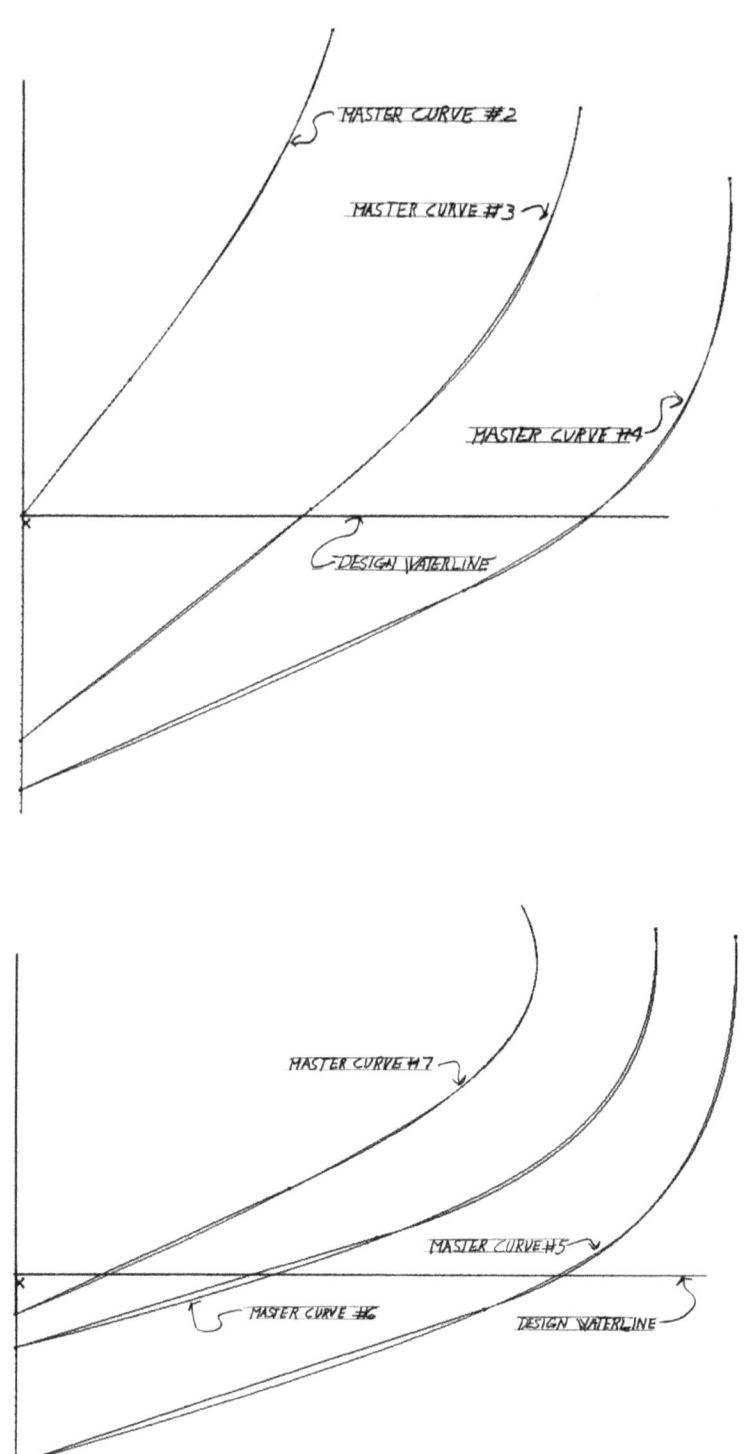

Double True Round surface hull designs:

A Double Surfaced hull designs combine both an Elementary and Warped surfaces into a single hull design. The design shown is the 'Prototype' and 'Proof of Concept' build using Approximate Development.

Here the upper surface is a True Round surface that begins at the sheerline, continues around the turn of the bilge and ends seamlessly and tangent at the theoretical chine line. The theoretical chine line is shown in the Section drawing show below.

A developable surface, then picks up seamlessly and tangent from that theoretical chine line to the centerline at the bottom of the hull.

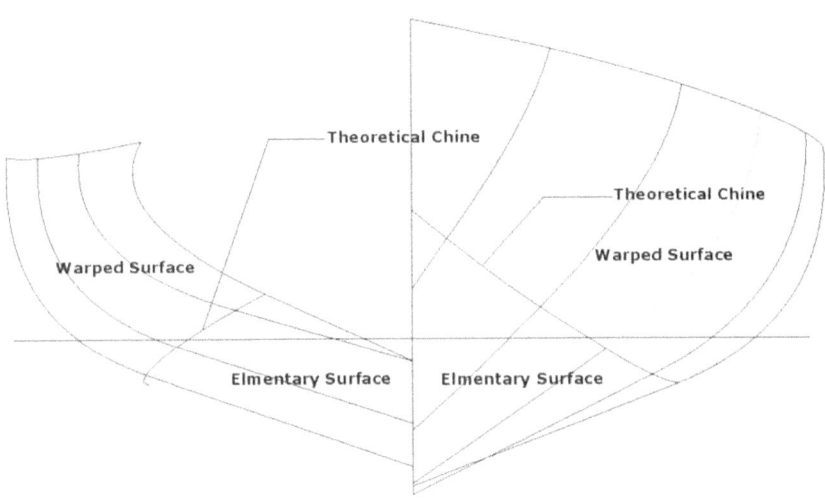

Triple surface True Round hull designs:

This hull configuration combine two Elementary surfaces and a Warped surface to define the hull. Here the upper surface is an 'Elementary' surface defining the freeboard of the hull. It begins at the sheerline and ends seamlessly and tangent to a Warped that defines the center surface at the turn of the bilge, which in turn ends tangent to another Elementary surface that defines the bottom surface of the hull.

In the 'Section' drawing, below, I have labeled the free-form curves at Frames Three, Seven, and Eleven with an approximate radius to demonstrate why this surface is indeed True Round. The drawing showing how the hull changes radius along the length of the hull.

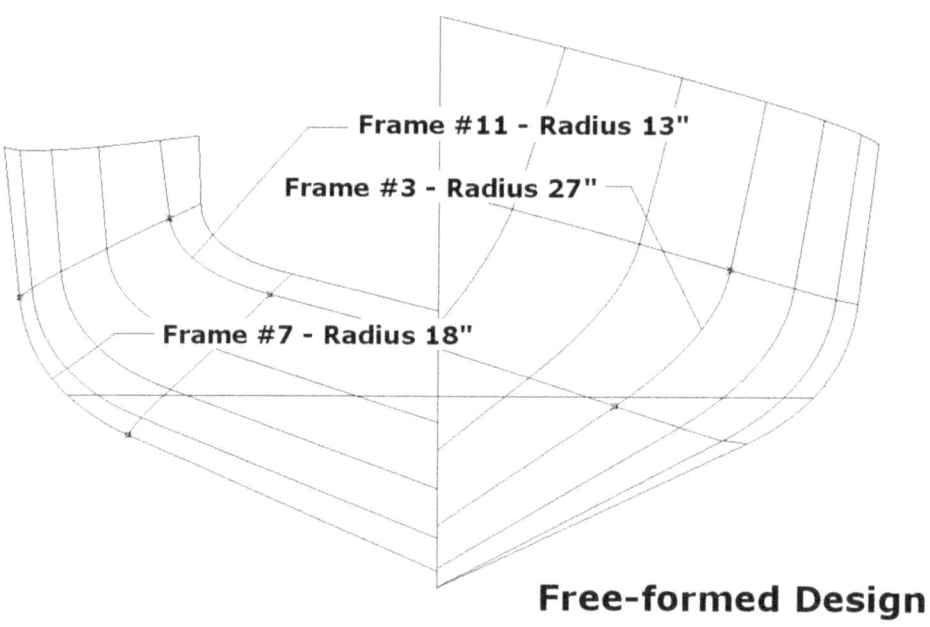

Free-formed Design

Triple surface Radius Chine hull designs:

The middle surface could be designed with a 'Single Constant radius' to becoming a 'Radius Chine Design', however using the design criteria for a 'Radius Chine' design, a totally different hull form would emerge as seen below. An eight inch(8") and sixteen inch (16") Single Constant Radius Chine show below were arbitrarily chosen. The Designer can use any radius, but the shape and hydrostatic will certainly change.

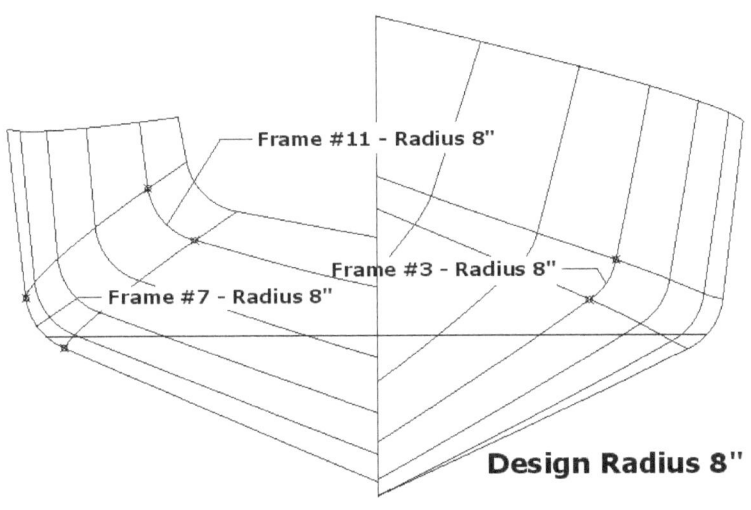

Design Radius 8"

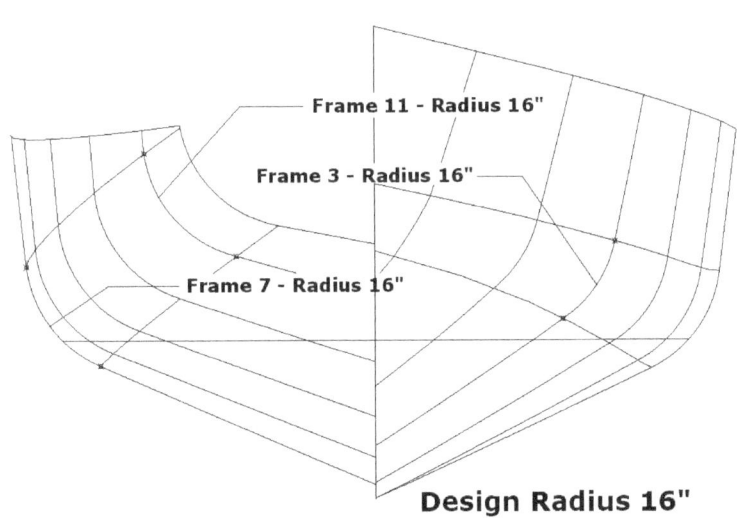

Design Radius 16"

Notes

Theory of Approximate Development

In sheet-metal work, surfaces are divided into two general classes: 'Elementary' and 'Warped'.

Elementary Surfaces:

Elementary surfaces can be developed accurately onto a single plane and formed back to their original three-dimensional form by simple folding and rolling techniques without stretching or compressing the material.

Elementary Surfaces are surfaces that:

- Lie on a Single Plane
- Curved Surfaces or sections of a curved surfaces.
- Radial Surfaces or sections of a radial surfaces.

An example of a Plane Surface is a rectangular box. Here random lines can be drawn anywhere on any side of the box. Lines such as these are designated as 'Elements' of the surface, because they lie in full contact with the surface.

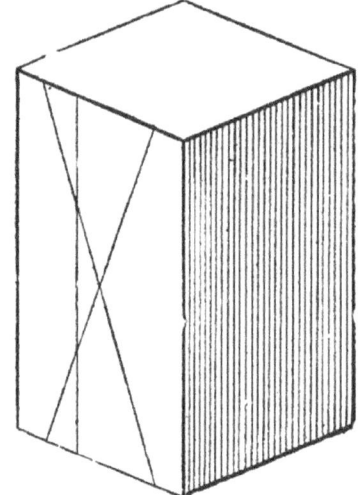

A Curved surface is a surface where no three consecutive parallel 'Elements' are on the same plane. To be an 'Elements' of a curved surface it is required to be in full contact with the surface.

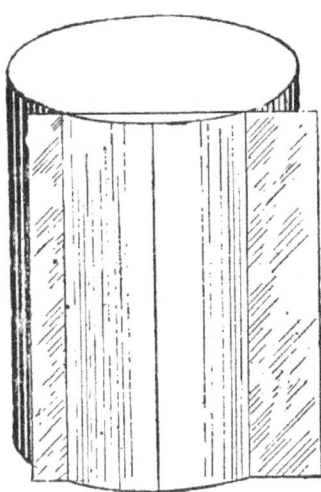

A Radial surface is again a surface where no three consecutive radial 'Elements' lie in the same plane. To be an 'Elements' of a surface radiating from an apex it is required to be in full contact with the surface.

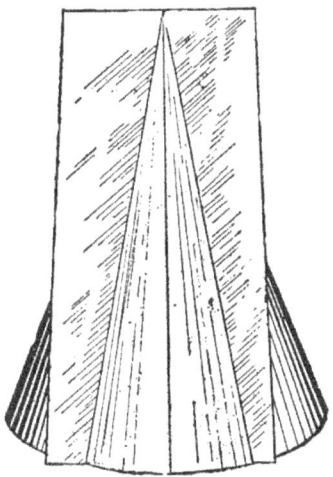

Any Sheetmetal or Plate fabrication composed of any combinations of these surfaces can be developed accurately onto a single plane. This includes such 'Fabrications' as Single, Double, and Multi-Chined metal and plywood boat hulls.

For sheet-metal and plate objects consisting of all Plane surfaces mathematical formula's such as **'Bend Allowance'** *and* **'Bend Deduction'** are used to 'Unfold' the fabrication onto a single Plane. Full details can be found in my book, ***'Applied Metal Boatbuilding Methods - Sheetmetal Pattern Development'.***

For sheet-metal and plate objects that are a combination of Elementary Planes, Curves, and Radial 'Elements' layout methods know as Triangulation, Parallel Line development, and Radial Line development are used to 'Unfold' the fabrication onto a single Plane.

Warped Surfaces

A warped surface has no 'Elements'. It is a surface where a straight edge makes contact at a single point only. A Sphere is a good example of a surface that cannot be 'Unfolded' 100% accurately onto a single plane. However, such a surface can be developed 'Approximately' to lie on a single plane.

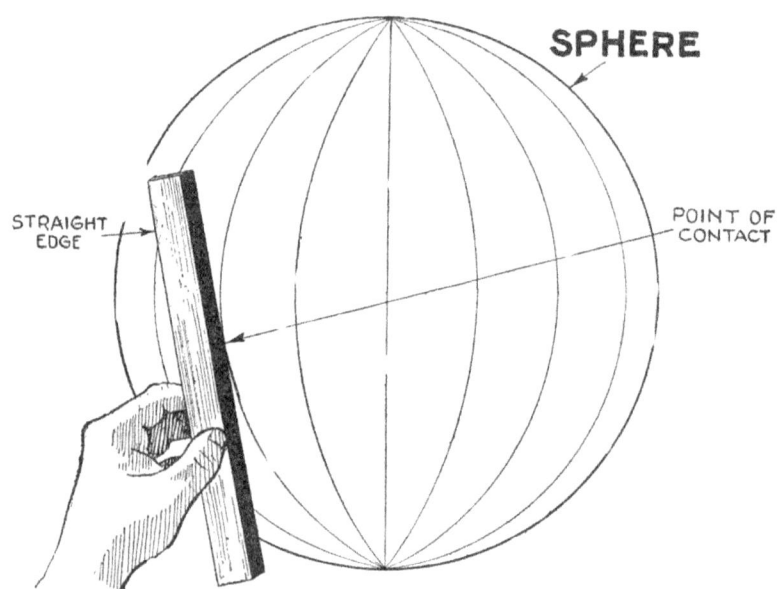

The Sphere is a good place to begin to understand how 'Approximate' development works, since it is a symmetrical object where only one pattern is required as compared to the many patterns required to 'Unfold' the warped surface of a true round metal boat hull.

Since 'Approximate' can be a relative term, a plan view of two (2) twelve-inch (12") spheres are shown cut at their diameters. One Sphere will be divided into eight (8) chords-segments and the other into sixteen (16) chords-segments. The chord line represent the segment of the Sheetmetal sphere developed 'Approximately' onto a single plane.

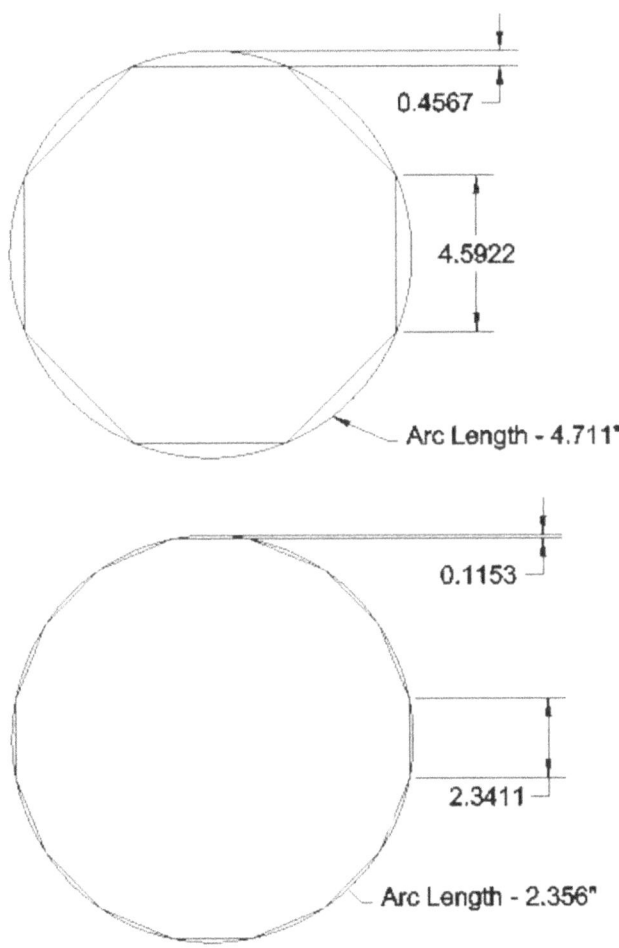

We could have divided the sphere into even more chords-segments, say thirty-two (32) or even sixty-four (64) or any amount we desire. Never the less the eight (8) and sixteen (16) chord arrangements will adequately describe the theory of 'Approximate' development for Warped or Compound surfaces.

illustrates the Eight (8) 'Approximately' developed chords-segment layout 'Unfolded' onto a single plane using 'Parallel Line Development'.

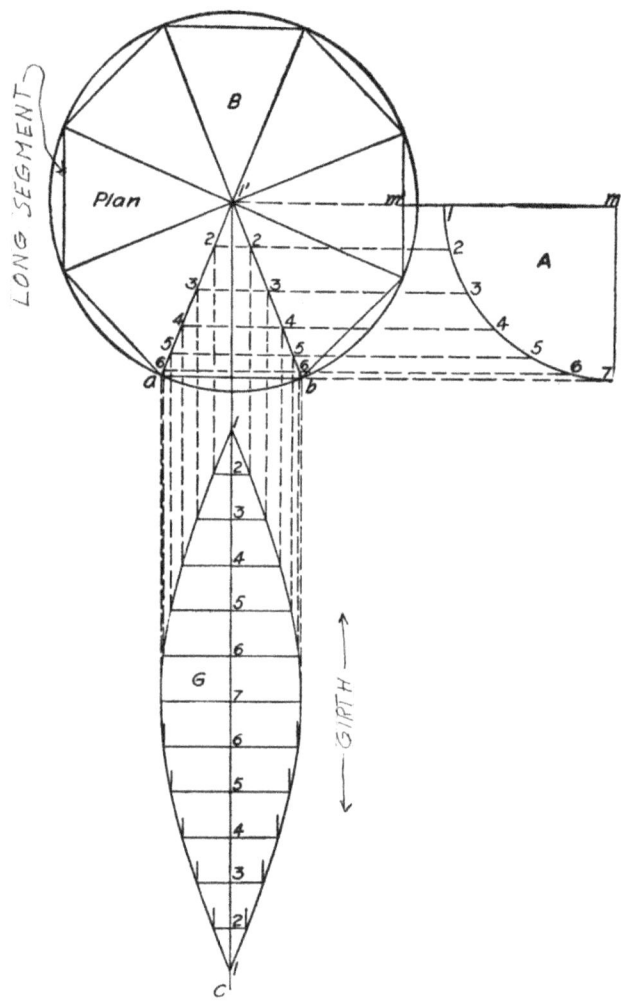

In the eight (8) segment layout, the chord length is 4.592", the distance between the chord and the finished surface arc of the sphere is 0.456", and the arc length of the finished surface of the sphere is 4.711". The difference in length between the chord length and the arc length is 0.199". The sphere therefore has been 'Approximately' developed.

When assembled the eight (8) segment sphere would be very prismatic in form.

The sixteen (16) segment sphere, however, would be less prismatic in form. In the sixteen (16) segment layout the chord length is 2.341", the distance between the chord and the finished surface arc of the sphere is 0.115", while the arc length at the finish surface of the sphere is 2.356". The difference in length between the chord length and the arc length is now 0.015".

The 'End User' could accept the eight-sided sphere for the top of their roof final or they could opt for the sixteen-side version which brings the fabrication closer to the true surface of the sphere.

If neither of the above were acceptable a peening or hammering process would be required to stretch the material to its true shape as seen below. It should be apparent that the eight (8) segment version would require more peening, but have less seams, while the (16) segment version would require less peening, than again there are more seams.

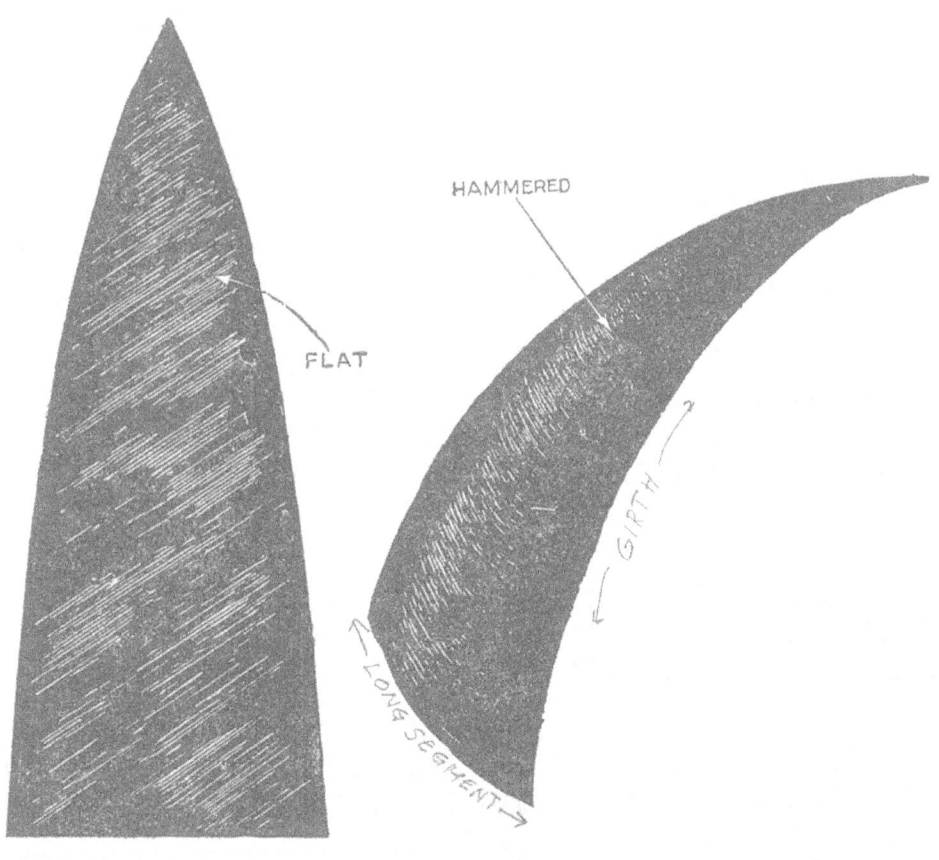

Approximate Development of Shell Plating:

Before proceeding to 'Unfolding' the Warped surfaces of a True Round Metal boat hull, we need to relate the terms 'Girth' and 'Length' used in hull design to the Sphere below to relate the two vastly different 'Fabrications' to each other.

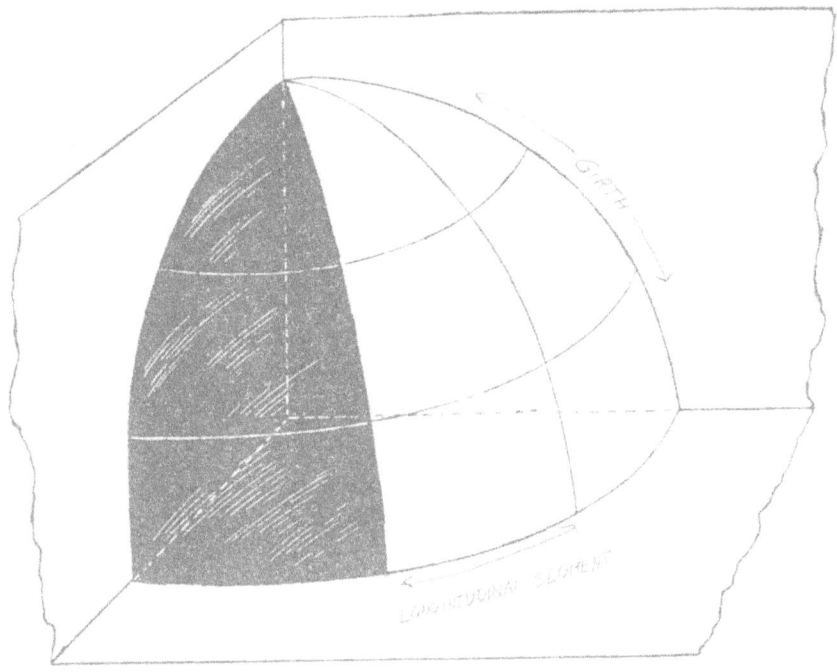

The only real difference between the two objects is that the Sphere needs only a single 'Approximately' developed pattern to define every segment of the sphere in both 'Girth' and 'Length'. Whereas the numerous segments that makeup the surface of a true round hull need to be developed individually since the 'Girth' and 'Length' vary at every longitudinal segment of the hulls shell plating.

Below the surface of the Bezier 12.5 has been divided into twenty-three (23) sub-surface segments along the length of the hull, which corresponds to the segments use to divide the Sphere.

For clarity, only the sub-surface segments that coincide with a transverse frame are shown. There are however two sub-surface plating segments between the one's shown in the below illustration.

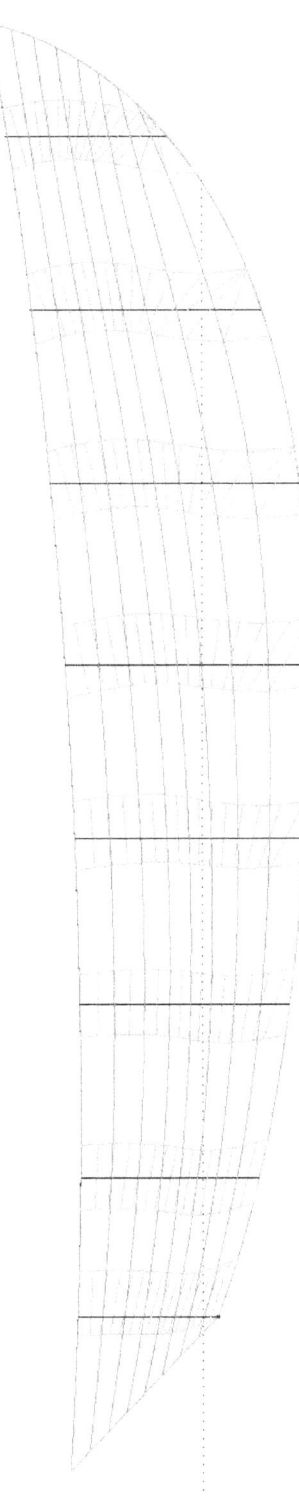

Longitudinal Direction of Roll:

Just as the Sphere was divided into chords at its diameter the free-formed curve of the Sheer-Line, shown in 'Plan View', is used to illustrate the chords along the length of the hull. The freeform curve represents the true form of the 'sheer-line', while the chord segments represent the 'Approximately' developed prismatic 'Sheer-line' intrinsic to 'Approximate Metal Fabrication'. The nodes define each segmented surface.

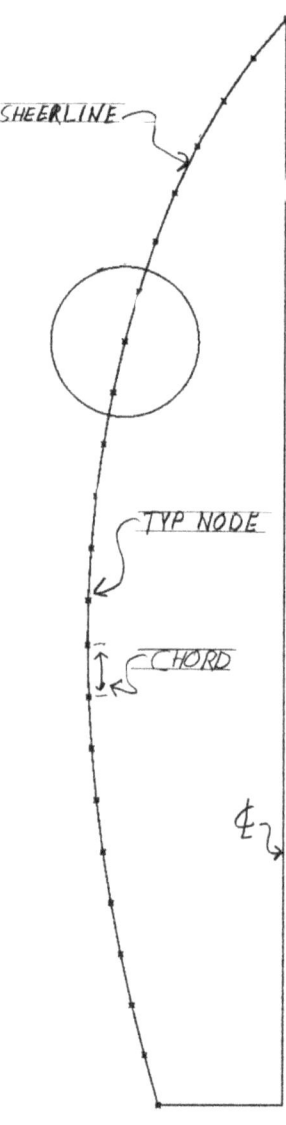

The below drawing illustrates a single line view between the free-formed sheerline and the chorded sheerline. Here the length of the chord is 8.681", while the length of the free-formed sheerline between the chords endpoints is 8.687". The difference being 0.006". The distance at the centerline of the chord to the free-formed 'Sheer-line' is 0.047".

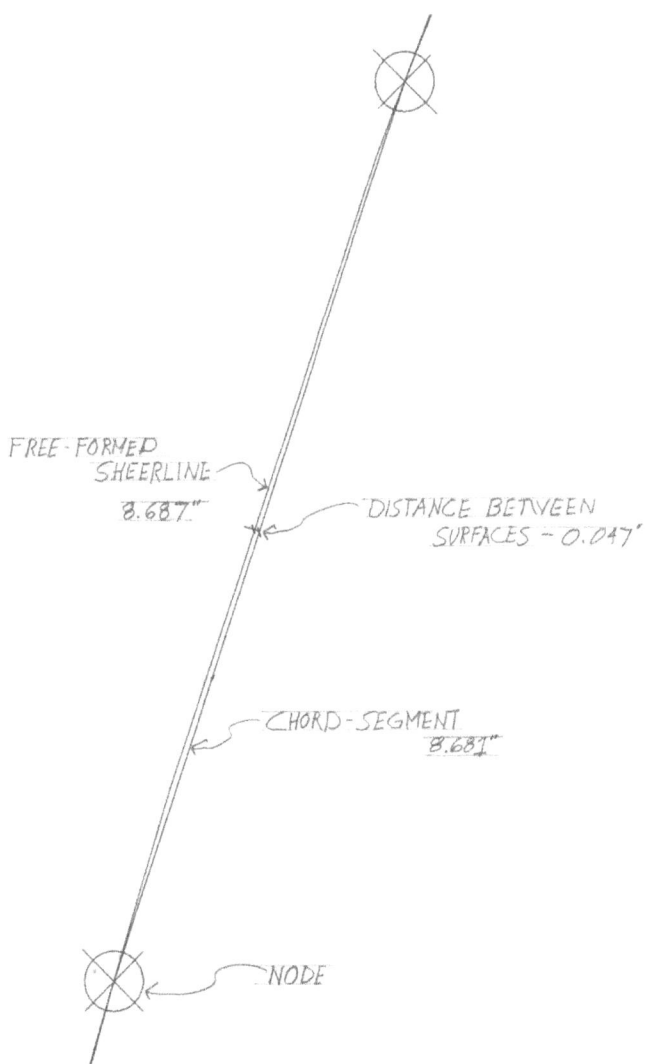

A secondary mechanical fairing procedures to reduce the difference between the true curve of the surface and the approximately developed chord surface would more than likely introduce unfairness to the surface.

It would seem to me that fairing compound would be the only solution, if this difference caused a surface fairing issue.

The following overall illustration, at the sheer-line, shows the relationship between the Chorded shell plates which were calculate by the process of 'Approximate Development' to the Compound curved surface from which the chorded shell plates were derived.

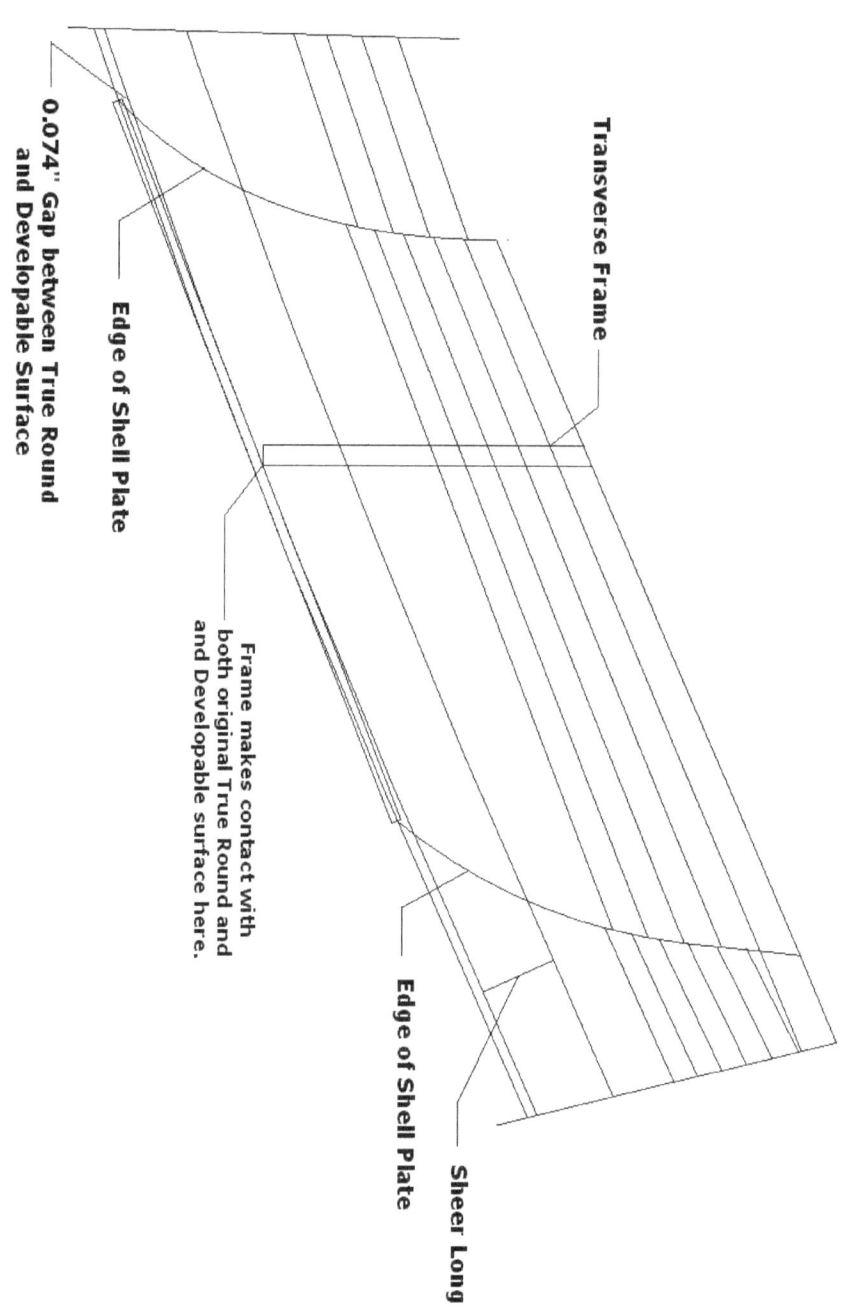

The following detailed illustration shows a closeup view of one-half of the above Chord segment. Notice that the segmented surface shell plate, unsegmented hull surface, transverse frame, and longitudinal all fall at a single location.

Also notice, while the designed True Round hull surface and the longitudinal frame both curve inward, while the segmented shell plate surface run straight out.

Here is where the 'Approximate' in 'Approximate Development' revels itself at the ends of the Chorded segment. In this case there will be a 0.074' standoff between the longitudinal frame and the chord segment shell plate that needs to be drawn together.

For more details on fabrication of the Longitudinal components and assembly of the hull refer to my other book
'True Round Metal Boat Building'

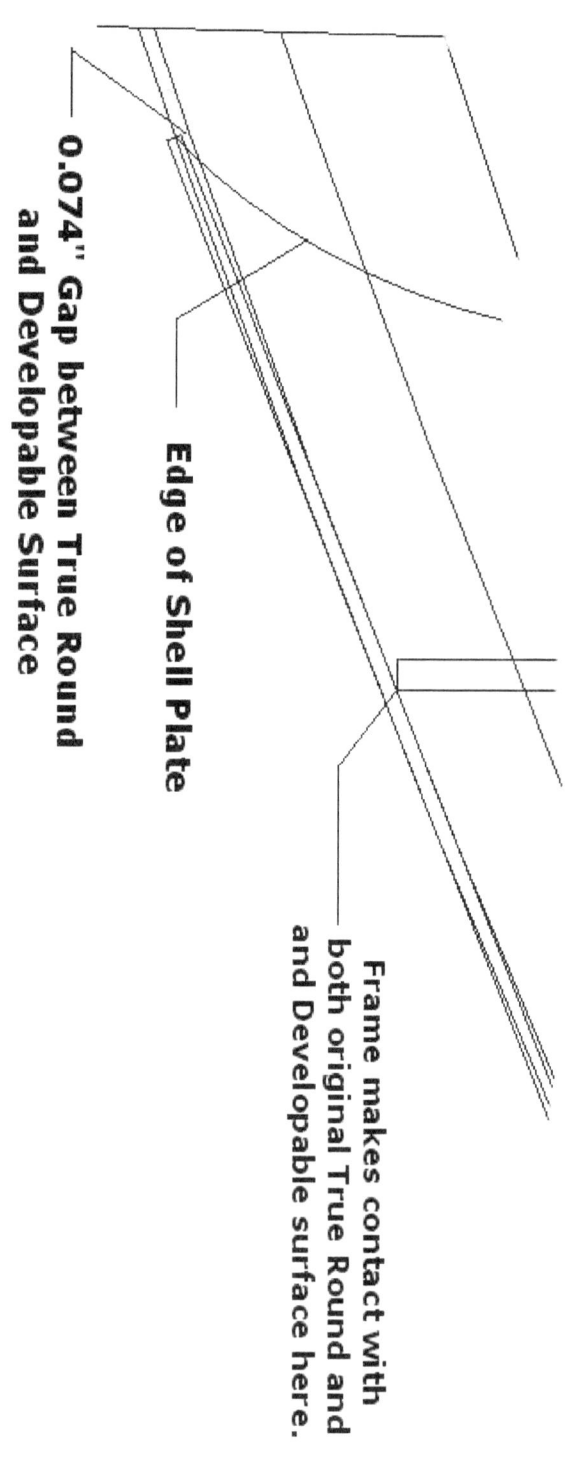

The Roll in Girth:

Dividing the length of the hull into segmented surface Chords is one direction of roll. Dividing those segmented surfaces into chords become the other direction of roll.

A 3D view of the segmented surface at Transverse Frame Three is illustrated below. The horizonal or nearly horizonal line represent three-degree (3) Bend Lines. They are the 'Elements' of the surface making the space between these lines, Chords.

Notes

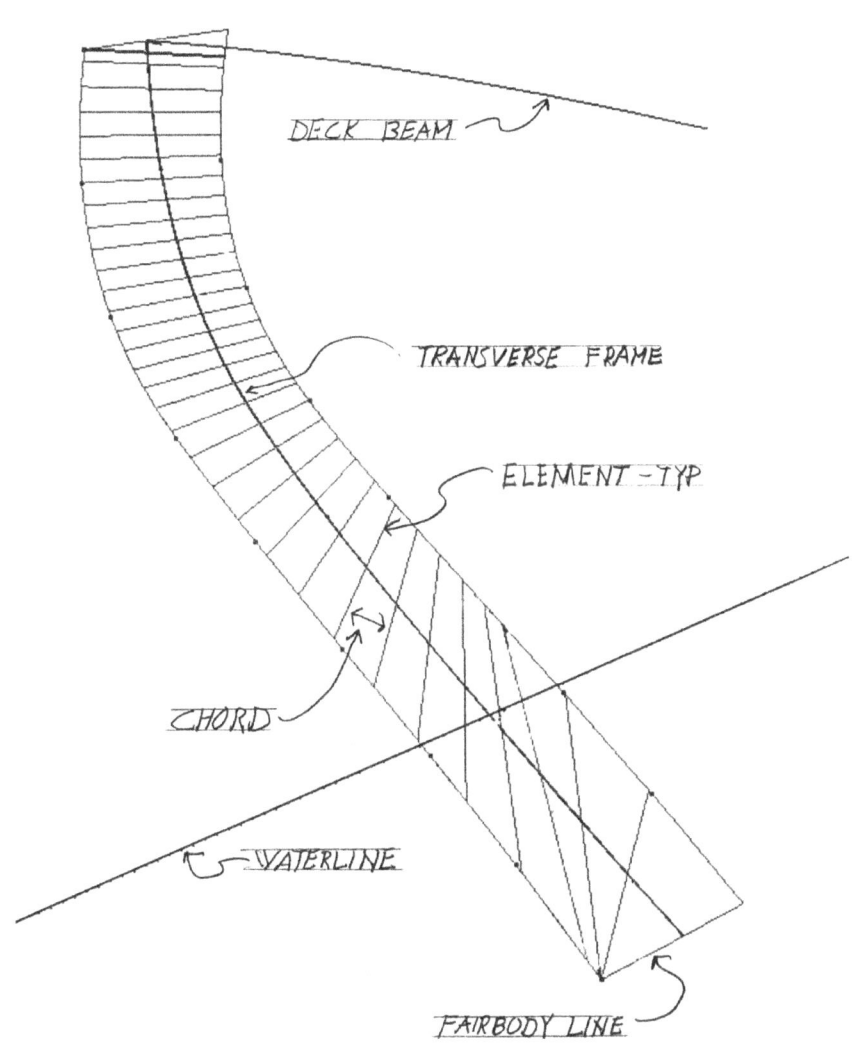

Notes

Hierarchy of Entities:

In this book I used Multi-surf by Aerohydro, a three-Dimensional **Surface** marine design software Package based on a 'Hierarchy of Entities' where complex surface designs are built up from 'Points', to 'Curves', to 'Beads', to 'surfaces', to 'Snakes', to' Ring' Objects, in the form of 'Parent-Child' relationships.

and

A 2D dimensional Cad drawing program, where Architectural drawings, Full size pattern, and CNC cutting files can be defined.

Points - Curves - Beads:

In this example there are four (4) **(Absolute Point)** objects named *control pt #1, control pt #2, control pt #3,* and *control pt #4*. These points will be used to define a **(B-Spline Curve)** named **base curve.** The 'points' are the parents of the 'curve'.

With 'Parent'-Child' relationships you cannot delete any of the 'points' that define the 'curve', without deleting the 'curve' first. The 'curve' **base curve** is the 'Child' of the 'Parent' 'points':

Control pt1, control pt #2, control pt #3, and control pt #4. However, you can move a point to another location and the curve **base curve** will instantly update to reflect that change.

To locate a position on a 'Curve', an **(Absolute Bead)** would need to be created. The 'bead' would be constraint to lay on the chosen 'curve. Its position would be determined by the (Parameter 't'). The start of a 'curve' the (Parameter 't') location is always '0'. The end of a 'curve' would have a (Parameter 't') location is always '1'. A location beyond the start and end of the 'curve' will follow the nature extension of the 'curve'.

The (Parameter 't')' location of the 'bead', named, **bead on curve** is located at a 't' of 0.85. The 'x', 'y', and 'z' location of any 'bead' can be found by opening the 'Beads' Properties.

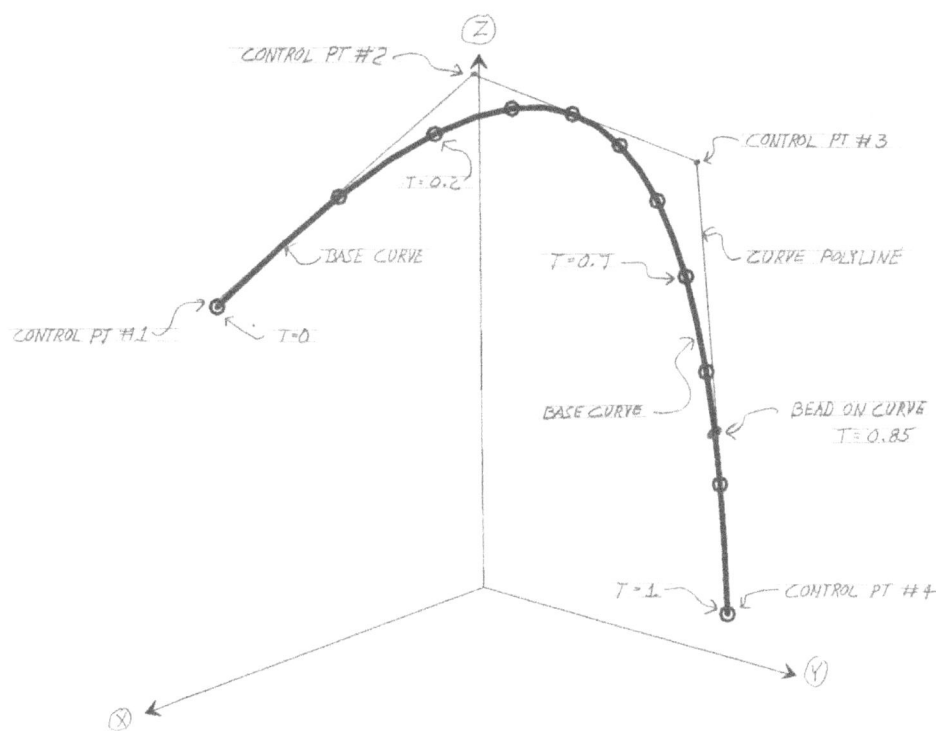

Surfaces - Magnets:

For clarity, the parent 'points' that define the (3) three **(B-Spline Curves)** in named **curve one, curve two,** and **curve three** have been hidden.

These three 'curves' support a **(C-Lofted Surface)** named *compound surface*. The three support curves are the parents of the **(C-Lofted Surface).**

To locate a position on a *'Surface'* a *'magnet'* is created. A *'magnet'* is constraint to lay on a 'surface'. Its position would be determined, again, by (Parameter 't'). There are two (2) values for (Parameter 't') for a 'surface' entity. They are 'u' and 'v'. In this case, (Parameter 'u')' runs longitudinally and (Parameter 'v') runs vertically. The four (4) corners would determine the values of (Parameter 't') at each corner of the 'surface'.

The (Parameter 't') location of the *'magnet'* named **position** has a (Parameter 'u' –'v' of 0.625,0.625. The 'x', 'y', and 'z' locations of any 'magnet' can be found by selecting and viewing its properties.

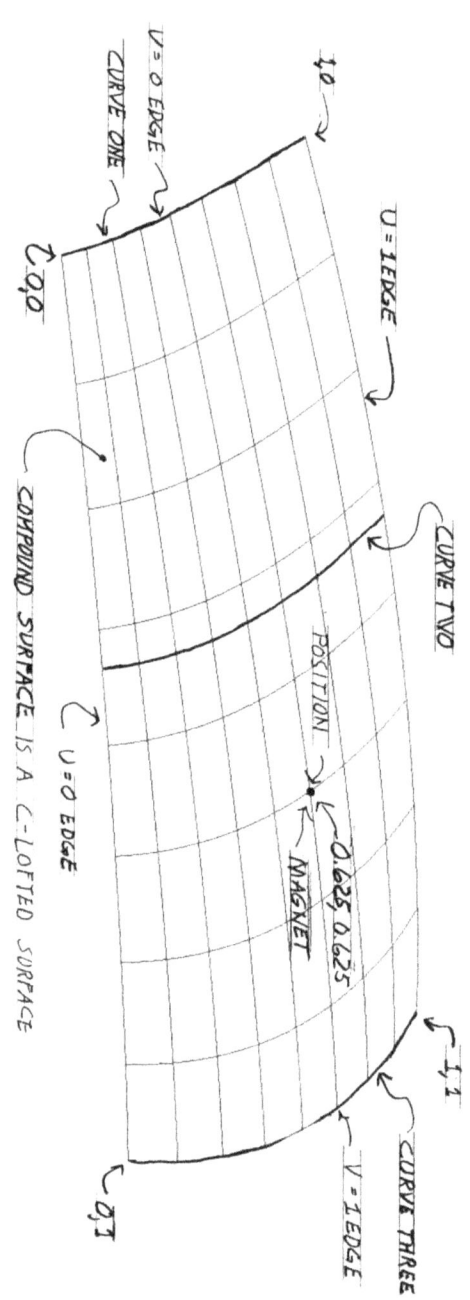

Snakes - Rings:

Snake one is a **(C-Spline Snake)** located on **compound surface**. It is defined by four (4) *'magnets'*: **mag one, mag two, mag three,** and **mag four** which are restraint to lie on **compound surface**. **Snake one** is the child of the four 'magnets'.

(Absolute Ring), *named* **ring one,** will be the 'child' of **snake one**. As with 'Curves' the value of (Parameter 't') is used to define location of a 'rings'.

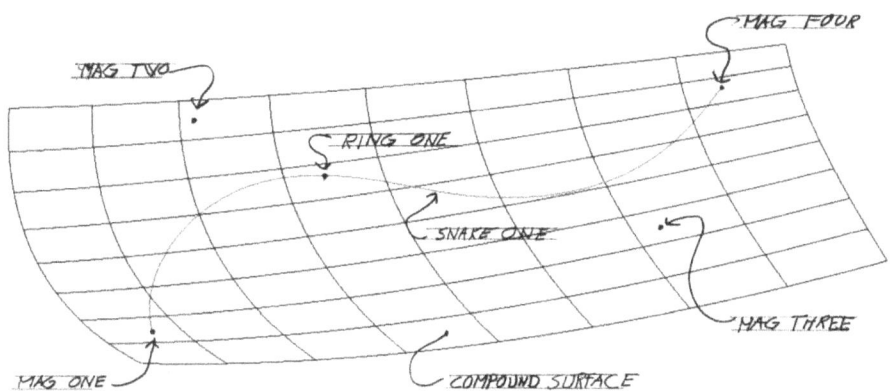

The Parent-Child relationship goes on and on. If the Designer decided to delete **snake one** for example, he could not be able to do so without deleting the *'ring'* named **ring one** first.

Ruled Surfaces

or

Developable Surfaces

Why my 'Bezier Design and Construction' will always use the 'Ruled' surface entity to define 'Longitudinal Frame' surfaces.

The following illustration, next page represents several 'unfolded' pattern for the sheerline longitudinal that is fabricated from 1.000" x 0.250" flat bar.

- The top drawing shows the 'unfolded' longitudinal frame that was defined by a 'ruled' surface. It was 'parented' by:
    - 'offset curve' – long_s_ht
    - 'Offset curve' – long_s_width

- The top-middle drawing shows the 'unfolded' longitudinal frame that was defined by a 'developable' surface. There is obviously a fatal error in the unfolding process possibly due to the twist in the forward end for this longintudinal surface, which is also 'parented' by:
    - 'offset curve' – long_s_ht
    - 'Offset curve' – long_s_width

- The bottom-middle drawing shows the 'unfolded' longitudinal frame divided into two distinct 'developable' surfaces based on 'sub curves' 'parented' by the aforementioned 'offset curves'.

- The bottom drawing shows an overlay of the top 'ruled' surface and the bottom-middle two-part 'developable' surface. As we can see, They overlay each other exactly. I have never a case where they did not.

Since a longintudinal frame surface defined by a 'ruled' surface and the two-part 'developable' surface overlay each other exactly, in all cases, my 'Bezier Design and Construction' designs will always use a 'ruled' surface entity to define longintudinal frames, since Multi-surf un-folding software 'Msdev will unfold 'ruled' surface onto a single plane, after a 'warning message', simplifying the design process.

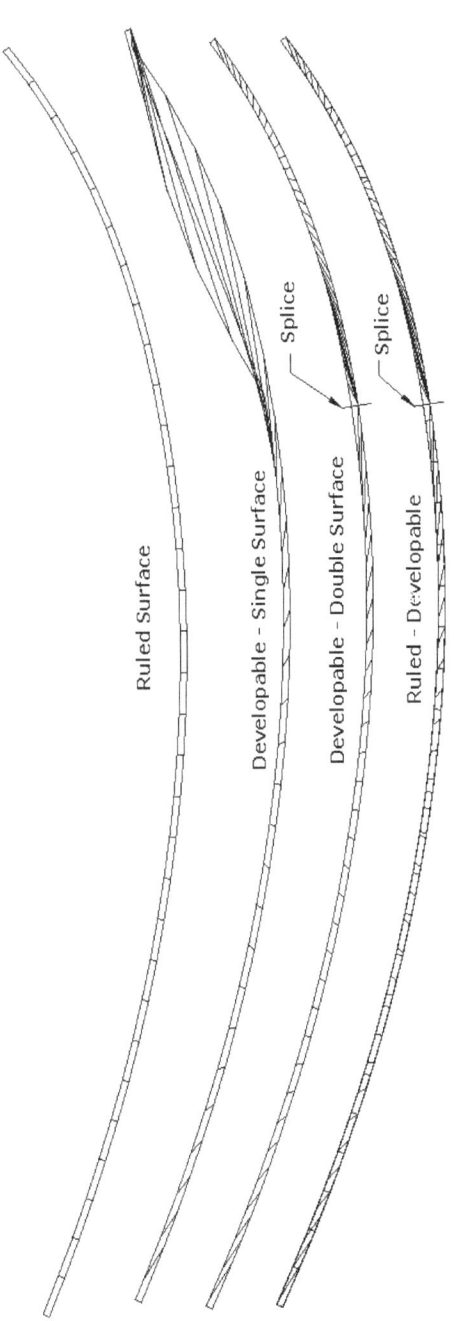

Notes

The Master Curve's Absolute Points

(The Absolute Points given below to create the Hull Model used in this tutorial are for instruction purposes only. They are only provided for the instruction use.)

Master Curve One

'abspoint' -   ***mc_1_pt_1***   'x' = 0.000"
              'y' = 0.000"
              'z' = 27.744"

'abspoint' -   ***mc_1_pt_2***   'x' = 5.322"
              'y' = 0.000"
              'x' = 14.887"

'abspoint' -   ***mc_1_pt_3***   'x' = 12.694"
              'y' = 0.000"
              'x' = 6.075"

Transverse Master Curve Two

'abspoint' - ***mc_2_pt_1***   'x' = 20.500"

    'y' = 15.840"

    'z' = 24.216"

'abspoint' - ***mc_2_pt_2***   'x' = 20.500"

    'y' = 13.960"

    'z' = 18.355"

'abspoint' - ***mc_2_pt_3***   'x' = 20.500"

    'y' = 7.057"

    'z' = 8.515"

'abspoint' - ***mc_2_pt_4***   'x' = 20.500"

    'y' = 0.000"

    'z' = -0.072"

Transverse Master Curve Three

'abspoint' - ***mc_3_pt_1***   'x' = 50.500"

    'y' = 28.356"

    'z' = 20.424"

'abspoint' - ***mc_3_pt_2***   'x' = 50.500"

    'y' = 27.242"

    'z' = 12.896"

'abspoint' - ***mc_3_pt_3***   'x' = 50.500"

    'y' = 18.014"

    'z' = 2.143"

'abspoint' - ***mc_3_pt_4***   'x' = 50.500"

    'y' = 0.000"

    'z' = -11.286"

Transverse Master Curve Four

'abspoint' - ***mc_4_pt_1***   'x' = 95.500"

    'y' = 35.988"

    'z' = 16.896"

'abspoint' - ***mc_4_pt_2***   'x' = 95.500"

    'y' = 36.341"

    'z' = 7.401"

'abspoint' - ***mc_4_pt_3***   'x' = 95.500"

    'y' = 26.556"

    'z' = -3.069"

'abspoint' - ***mc_4_pt_4***   'x' = 95.500"

    'y' =   0.000"

    'z' = -13.781"

Transverse Master Curve Five

'abspoint' - **mc_5_pt_1**   'x' = 140.500"
                                       'y' = 33.972"
                                       'z' = 15.756"

'abspoint' - **mc_5_pt_2**   'x' = 140.500"
                                       'y' = 34.168"
                                       'z' = 6.840"

'abspoint' - **mc_5_pt_3**   'x' = 140.500"
                                       'y' = 24.220"
                                       'z' = -2.233"

'abspoint' - **mc_5_pt_4**   'x' = 140.500"
                                       'y' = 0.000"
                                       'z' = -8.868"

Transverse Master Curve Six

'abspoint' - ***mc_6_pt_1***   'x' = 161.500"
                                                              'y' = 30.192"
                                                              'z' = 16.080"

'abspoint' - ***mc_6_pt_2***   'x' = 161.500"
                                                              'y' = 30.537"
                                                              'z' =   8.849"

'abspoint' - ***mc_6_pt_3***   'x' = 161.500"
                                                              'y' = 17.577"
                                                              'z' =   0.857"

'abspoint' - ***mc_6_pt_4***   'x' = 161.500"
                                                              'y' =   0.000"
                                                              'z' =   -3.432"

Transverse Master Curve Transom

    'abspoint' - **ap_sheerline**   'x' = 185.604"
                                                                 'y' = 23.832"
                                                                 'z' = 17.184"

    'abspoint' - **ap_bezier_guide**   'x' = 180.326"
                                                                  'y' = 26.500"
                                                                  'z' = 12.104"

    'abspoint' - **ap_chine**   'x' = 171.072"
                                                                  'y' = 12.608"
                                                                  'z' =   3.171"

    'abspoint' - **ap_fairbody**   'x' = 165.816"
                                                                   'y' =   0.000"
                                                                  ' z' =   -1.896"

Notes

The Bezier 12.5

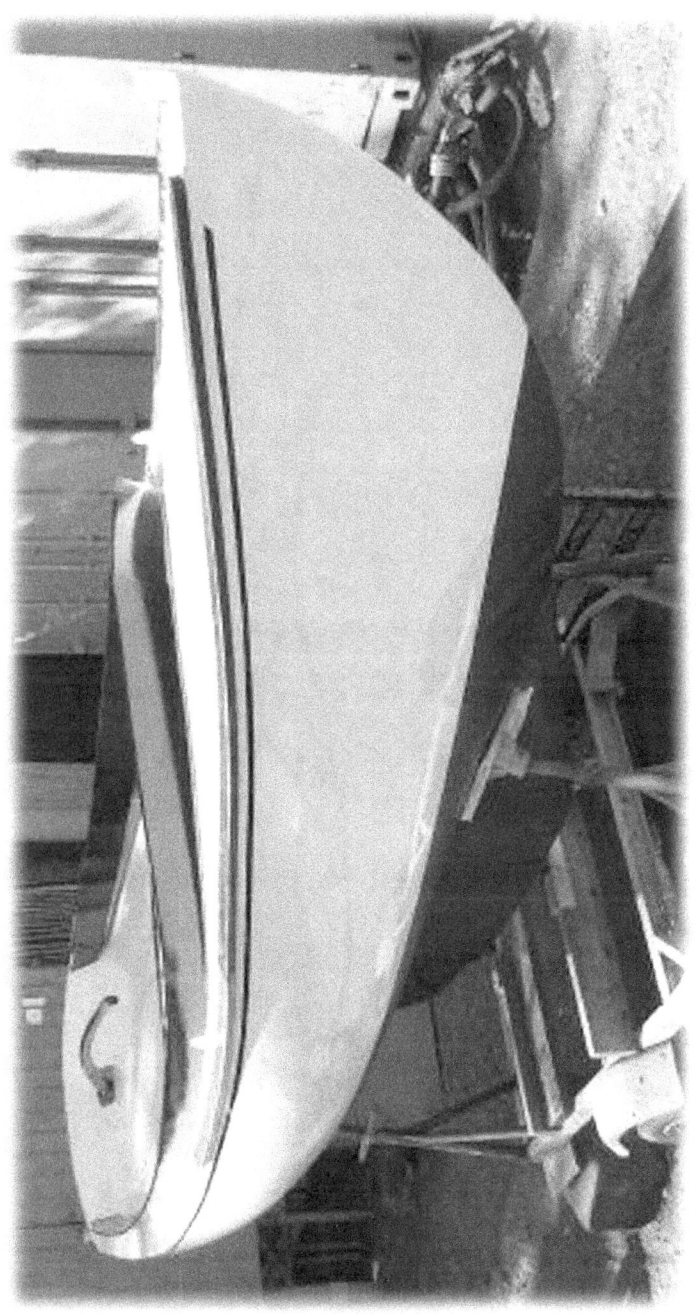

Other Bezier Books

at Amazon

Steel Mast Design and Fabrication
True Round Metal Boat Building
Fabrication of Hull Integrals

Notes